Alien Worlds

Biology, Society, and Spirituality Beyond Earth

Skye Alden

22 Lions

Alien Worlds: Biology, Society, and Spirituality Beyond Earth

Written by Skye Alden

Contents

Introduction

Imagine stepping beyond the sky's edge, where life takes forms that surprise and inspire. Alien Worlds: Biology, Society, and Spirituality Beyond Earth invites you to meet beings and societies formed by worlds nothing like our own. Life there isn't just getting by—it moves with a rhythm of respect and balance that touches everything. You find creatures shaped by strange places and communities built on cooperation and shared goals.

In these stories, technology walks alongside nature. Progress doesn't fight the world; it works with it. These worlds show life supporting life, each part lifting the others. Culture, art, and music bloom in ways that stretch beyond what we expect. Spirituality here flows from love and connection, not from fear or control. These alien tales ask us to see ourselves differently, to rethink how we live and what being alive really means.

If you wonder about what might be out there or want a fresh way to look at life, this book offers a gentle burst of wonder. It calls you to open your heart and mind, to imagine beyond the usual. Alien Worlds isn't just about life beyond Earth—it invites you to widen your view of life itself.

Chapter 1: Introduction to Extraterrestrial Life

Since the dawn of human thought, we have glanced upward at the night sky and felt something stir within us—a question more alive than mere curiosity. Is there life out there? Beyond our small planet, somewhere on a distant dot circling a fading sun, might beings like us exist? This question has seeded the minds and hearts of many, crossing the borders of science, myth, and spirit. For some, it's not just an idea but a memory or a feeling, a whisper of contact with something beyond Earth's bounds.

History holds voices who spoke of such encounters. Their words may have been brushed aside or caught in the winds of skepticism, yet they widen a door—just enough for us to peek through—to see how far life's branches might stretch in the vastness beyond our skies.

Take Emanuel Swedenborg in the 18th century. Here was a man who stood with one foot in the lab and the other in the world beyond the seen. Through his visions, he painted places not unlike ours yet brimming with a shared wisdom and peace Earth continues to seek. Swedenborg's otherworldly societies did not simply survive; they moved with a spirit of balance and clarity, as if echoing an unspoken truth hidden just below our awareness.

Jump forward to the mid-1900s—George Adamski enters the scene, creating waves among those who looked at the skies for answers. He claimed to have met visitors from Venus and other planets, describing beings who carried both warmth in their gaze and knowledge far deeper than any technology we had then—and maybe even now. These were not just travelers of space but messengers blending science with spirit.

Then there is Edgar Cayce, the "Sleeping Prophet," whose trance readings reached into the cosmic web. Cayce's message was not confined to physical realms but threaded through the unfolding dance of consciousness itself. He spoke of beings whose existence played a part in the grand unfolding of awareness across the universe—a reminder that life's story is far wider than flesh and bone.

A thread running through these tales is clear: life beyond Earth wears many faces. Some forms are pure energy, flickering and alive in unseen ways. Others appear humanoid, not because they mimic us, but because intelligence often finds similar shapes for communication and growth. Many possess qualities we do not. They survive where we cannot imagine—whether wrapped in thick clouds or floating in the emptiness of space. This diversity calls us to ask again, what does "living" really mean?

Adaptation is not merely survival on these worlds. It is growth, evolution, and flourishing under conditions so strange to us they stun and inspire. It pushes back against our narrow definitions and invites us to expand our minds beyond what we have been taught.

The societies these stories describe are often startling. Rivalries and power struggles seem like distant echoes, replaced by systems built on respect and mutual aid. Technology serves everyone, lifting burdens instead of stirring conflicts. Life moves with a rhythm gentle yet profound, woven by shared values rooted in love and cooperation. Here, community is not just a word but a lived experience.

Look closer, and you find culture alive with connection. Their lives flow in harmony with nature's cycles, art blossoms as a language of the heart and spirit, and clothing honors both function and beauty with respect for their environment. Food nourishes more than bodies; meals become rituals, ceremonies blending gratitude for the earth with the bond between beings. Eating is a sacred exchange—a moment that draws the community and planet closer.

Spirit pulses through all these elements. Belief in a shared consciousness summons rituals that expand awareness and encourage kindness. Their moral compass draws from love and the understanding that harm to one ripples through all. This perspective reshapes how we see existence itself—as a tightly bound fabric where each thread matters.

When seen as a whole, these stories do more than inform. They invite us to widen our view, to embrace life's vast palette and allow the universe's diversity to kindle new ways of seeing. They challenge what feels settled about biology, society, and spirit, hinting at possibilities rather than fixed rules.

As we send ships and probes into the night, seeking signs beyond our world, let these voices hold space in our hearts. The stars may not be silent or empty. Instead, they might be neighborhoods filled with others—neighbors reaching across distances, offering connection and growth. In that boundless expanse, we might find ourselves part of a great unfolding story, reaching out and touching itself across the canvas of space and time.

There is wonder here. There is call. There is the hope that we learn to listen as much as we look, to open not only our eyes but our souls, discovering not strangers but kin—other sparks in the vast fire of being.

Chapter 2: Biological Diversity and Adaptation

The universe stretches out before us, filled with countless stars and planets. Each world carries its own story, its own conditions, and perhaps, its own life. People who have experienced strange encounters or have visions of other worlds share tales that spark curiosity. They talk about life forms very different from what we know on Earth, able to live and grow in places that seem harsh or impossible. These stories shift how we think about life itself, showing that it can take many shapes and survive where we might never imagine.

Take Emanuel Swedenborg, a thinker who described his visions of life beyond our Earth. He told of beings that didn't all look human. Some seemed to be made of light, or energy flowing freely. His words suggest life doesn't need to follow Earth's rules, like carbon-based bodies or the need for air we breathe. Instead, life might appear in countless forms, whatever fits the world they call home.

George Adamski, known widely for his claims of meeting visitors from other planets, adds another layer to this picture. The beings he described from Venus weren't just physical; they were refined, with senses sharper than ours and ways to communicate without speech. Their bodies had grown strong and flexible to handle conditions very unlike Earth's—different air, gravity, and even radiation levels. These traits didn't come about randomly but shaped by their home environments. They show that life can stretch and bend itself, rising to meet challenges or opportunities no living thing here has seen.

Then there's Edgar Cayce, whose readings often reached into spiritual truths along with the physical. His idea was that life elsewhere doesn't just change on the outside. The spirit inside, the awareness and growth of consciousness, plays a role too. Cayce said that each planet's unique climate and conditions help guide both the bodies and souls of its inhabitants, pushing evolution toward new heights. This view paints a picture where physical form and spiritual growth walk hand in hand, each influencing the other.

When you hear these accounts, one thing becomes clear: life outside Earth has to be adaptable. Some beings may live where the air is thick and heavy, where we'd suffocate or crumble. Their breathing systems and bodies work differently, allowing them to exist comfortably. Others might not need air at all, floating in space, made of energy or forms we struggle to define. That life would not just survive but flourish in ways we might never guess.

This adaptability doesn't stop at the body. Different worlds bring different struggles, and civilizations evolve to meet them. Where radiation is strong, some beings might wear natural shields or carry defenses woven into their very cells. On worlds where food and water are scarce, societies learn clever ways to recycle and keep their environment healthy for generations. Culture, technology, and biology weave together, each shaped by what the planet demands.

All these stories from visionaries and contactees open up a vast field of wonder. They push us to rethink old ideas about what life must look like or how it must behave. They invite us to imagine far broader possibilities—that life, in any corner of the cosmos, might be as varied as the stars themselves. This understanding can ground us, reminding us of how life on our own planet has managed to survive and change through time.

To bring it all into view: descriptions of life beyond Earth offer more than strange tales. They expand our sense of what life can be. They show how flexible and strong life is, able to grow in ways that challenge what science often assumes. As we look deeper into space and ponder these accounts, they encourage us to stay curious and open. There is more to life than our textbooks contain—a vast, living universe waiting for us to explore, understand, and respect on many new levels.

The universe is teeming with chances for life to appear, shift, and unfold in forms we might never have thought possible. And as we listen to those who give us glimpses of these other worlds, we remember the strength and beauty of life itself. It is a force that moves beyond borders, beyond science and story, always changing, always becoming something new. That is a powerful truth to hold as we keep our eyes turned skyward, ready to meet what comes.

Chapter 3: Social Structures on Other Worlds

When people talk about other worlds and the ways beings live there, something curious happens. The images they paint feel both strange and somehow familiar. Those who claim to have met or learned about creatures from beyond Earth often describe social setups that don't match what we're used to. These accounts let us peek into lives shaped by very different planets, by values and growth paths unlike ours—whether those relate to tools and machines, spirit, or culture. Considering these stories does more than entertain our imagination. It moves us to reflect on ourselves; on the ways we arrange our societies here, and on what truly matters.

What stands out first is this: these extra-earthly groups seem to build their lives around working with each other, not against. On many worlds, the drive to control or dominate fades into the background. We hear about communities where fairness and shared care live right at the surface. The people there don't act from competition or self-interest alone. Instead, they seem connected by a deeper kind of understanding, a feeling of togetherness that shifts attention away from the individual toward the whole group. Their thinking and being spring from a level of awareness that runs much deeper than basic survival or fighting for position.

Many accounts tell us these civilizations have advanced beyond what we call technology, but not merely in the sense of new gadgets or machines. True advancement comes from how they live together. Education, for example, is something they treat with great care. Children don't just fill their minds with facts. They nourish their spirits alongside their intellect. This way of learning helps people find what makes them special. It encourages everyone to bring their own unique gifts back to the community. The goal isn't to pull people apart but to weave them together, making a strong and living whole where the individual and the group both grow.

Another idea often mentioned is that beings from these places don't feel trapped by the old human ideas about gender and family roles. The rigid boxes we know don't apply there. Connection flows from respect and spirit, not from biology or set social rules. Such freedom builds a culture where self-expression can take many forms. The ways they relate reach beyond our expectations, revealing a mindset that values what exists between people as much as what is inside them. Fixed roles and old categories lose their hold in these stories.

Respect for their home planet is a vital thread in many of these tales. The societies described live gently upon their world, not by fighting nature but by moving alongside it. Their touch on the land is light. They care deeply about keeping the environment whole and balanced. Communities are organized with this reverence in mind, honoring the earth and all its creatures. Technology serves nature rather than standing apart from it. They use only what they need, leaving no waste behind. This respect shows a profound union where the social life and the world are parts of one large, living flow.

Adaptability shines through the descriptions too. The worlds inhabited by these beings can be harsh, with strange climates and elements that challenge life as we know it. So their social systems bend and flex to fit those conditions. The people there don't merely hold on, they flourish amid change. Their structures breathe with life, shifting when needed and welcoming the flow of the unexpected as part of existence.

These stories, told by those who say they have witnessed these civilizations, stretch what we think is possible. They move past our usual ideas about power, control, and order. Instead, they show lives built on cooperation, respect, and a close bond with nature. When we lift our eyes to the stars and wonder about life beyond Earth, these glimpses offer something hopeful and fresh. They tell us that life might take many paths, changing and growing in ways we have scarcely dreamed.

Looking into the social ways of beings from other worlds is more than curious guessing. It calls us to think about how we live here and what lessons we might borrow. Imagine how different our communities could become if we embraced ideas of mutual care and balance on a larger scale. What would happen if we opened ourselves to a wider sense of connection and support?

In the end, the tales of societies beyond Earth are not just fancy stories or dreams. They carry messages, questions, and open doors. They invite us to step beyond the limits that bind our thinking, to open up to the simple beauty of different ways life can find harmony and thrive. As humanity reaches outward to explore space, these ideas whisper a quiet truth: life is wonderfully varied and endlessly able to change. The future of how we live together can be brighter and more connected than we might have dared to hope just now.

Chapter 4: Cultural Systems and Lifestyles

Imagine stepping beyond the usual stories we hear about other worlds and their inhabitants. What if those beings, so often painted as strange or distant, live in ways that speak to something deeper and more grounded? When you listen to the experiences of those who believe they have made contact—those visionaries and travelers between worlds—you catch glimpses of lives shaped by balance and meaning, far from the rush of Earth's busiest cities.

Picture their days not as a scramble but as a dance. Instead of rushing around, many accounts share a different beat—one closely tied to natural flows, the sun's path, the phases of the moon, and the gentle turn of seasons. These rhythms shape everything from work to rest, from how meals are prepared to the way their communities gather. The ordinary becomes sacred. Cooking is a moment of presence, art isn't just decoration, but a language of feelings and spirit. Even technology melds with this mindset, crafted as an extension of mindfulness instead of mindless progress.

Community is the heartbeat in many of these stories. There's a strong sense of belonging and shared care, a feeling that every person's well-being matters deeply. Gatherings are frequent, whether to honor the stars, mark a harvest, or celebrate a collective achievement. People come together with music and dance, not just to entertain, but to open the soul and remind everyone of the ties that bind. These events offer reflection and learning, moments where life slows just enough for deeper insight.

Art and music hold a very special place. Imagine art that flows beyond the canvas or sculpture—art that plays with light and sound and even thought. Sometimes, creative works aren't just seen or heard but felt, as if the mind and heart are drawn closer through these expressions. Music in these societies can be complex yet soothing, carrying layers of feeling and spirituality. Creativity is often a shared process, a way to connect rather than separate, weaving individual gifts into something larger, something alive.

Learning builds the foundation of growth. From childhood, people in these cultures are encouraged to open every part of themselves—the mind, the heart, the spirit. Education is continuous, a lifelong journey, not a checklist or a race to finish first. Every day brings a chance to understand more, not just about facts but about the vast universe and one's own place within it. Knowledge serves the community and the unfolding story of life itself, not just the individual.

Respect for the world beneath their feet is evident. Stories from contactees often describe beings who move gently on their planets, nurturing the land instead of taking too much. Their farming blends with nature's patterns, producing food that feeds without harm. Technology reflects this care, designed to support life, never to destroy it. There's an understanding here—that life, all life, grows from a network of relationships, delicate and interwoven.

These visions don't just paint pictures of alien worlds. They hold up a mirror to us here on Earth. They nudge us to rethink what culture can be when it grows from mindfulness, connection, and respect for life's rhythms. As we look outward to stars and possibilities, these stories also turn our gaze inward, inviting change. If other civilizations can find harmony and meaning on their planets, why not us?

The universe, seen through these experiences, is a place rich with ways to live deeply. It calls us to remember that progress doesn't require destruction, that community is strength, and that every routine holds potential for grace. When we consider these ideas, the vast unknown feels a little closer, and our own journey toward a balanced life takes on new hope.

There is a quiet power in these glimpses—a gentle urging that life can be more connected, more thoughtful, more whole. These accounts speak quietly but strongly, reminding us that beyond the stars may lie not just other beings but other ways of being. Ways that invite us all to listen, learn, and grow beyond the edge of what we thought possible.

Chapter 5: Clothing and Fashion

Clothing and fashion among beings from other worlds paint a picture quite unlike our everyday experience. Those who claim contact with extraterrestrial civilizations often share stories that reveal how these beings dress not just for function, but as part of a deeper expression of who they are. It's a lively mix of culture, environment, and spirituality all woven into the fabric of their attire. What we might see as simple clothing, they understand as a language, a symbol, even a bridge to their surroundings. This invites us to expand our minds about what clothing can really be.

One thing that draws attention right away is the variety of materials these beings use. Earth's fabrics come mainly from animals, plants, and manmade fibers. Out there, it's different. Some garments are said to be made of crystal-like substances that shimmer like gems. Others shine with glowing threads or even flow as pure energy, shifting colors and shapes in ways that feel alive. These aren't just pretty to look at; they are designed to help with things like protecting the wearer from harsh climates, keeping an ideal temperature on body surfaces, or even syncing with gadgets they carry. The materials suit their worlds perfectly, blending beauty with practical needs.

The way these clothes are styled tells stories too. Often, the pieces can alter their form, growing longer or shorter, tighter or looser, as the wearer needs. Imagine a jacket that tightens to shield against radiation, then relaxes into light robes when the danger passes. The secret here seems to be their ability to rearrange themselves on the smallest level, where atoms and molecules move like dancers taking directions from a master. Some say these garments even breathe, grow, or repair themselves—almost like living things that walk alongside their wearers.

Colors run wild in their wardrobes. Bright, shifting hues splash and whirl, sometimes echoing the moods of the person inside the clothing. Geometric patterns pulse with meaning, their shapes not random but representing ideas, clans, or emotions. Sometimes images float like holograms across the fabric, telling stories or sharing feelings without words. Wearing these is like carrying a personal billboard, speaking silently but powerfully to others in the community. Social rank, family ties, and hidden thoughts become visible in these lively displays.

Their clothes hold more than just looks and uses; many of them carry a spiritual weight. Sacred symbols, energetic flows, or even tiny lifeforms embedded in the material bring a deeper purpose to the act of dressing. Wearing such attire might be a ritual, a way to stay in touch with the cosmos or tap into spiritual guidance. These garments are not just protection for the body—they guard the soul, nurture awareness, and connect the wearer to something greater than themselves. Dressing becomes a daily practice of respect and mindfulness.

Making these clothes is different as well. Instead of a factory line or a commercial store, it's often a shared, collective process. Community members come together, each bringing skills, stories, and prayers. The act of creation turns into a form of art and devotion. Every stitch or bonding of fiber carries intention, honoring the bonds between life forms and the universe. It's as much a celebration of connection as it is about making something to wear.

Thinking about these ideas opens a door. It pushes us beyond the typical view of fashion as just fabric and design. The tales from others who walk the stars reveal a far richer practice—one where the clothes we wear can carry meaning, power, and soul alongside beauty and practicality. They ask us to look at our own clothing not just as something to cover ourselves, but as a living reflection of who we are and how we relate to the world around us.

As explorers of the unknown, each new detail about these distant cultures reminds us how diverse life can be. But it also invites something close to home—a call to honor the earth, each other, and the mystery that ties us all together. Perhaps if we listen to these distant whispers, we might find ways to dress with more care, to create and wear garments that breathe with meaning and hold space for the sacred in everyday life. Clothing, then, becomes not just a layer on skin, but a thread woven through the ongoing story of existence.

The universe is wide and rich, full of stories waiting to be known. When we open our eyes and hearts to what others wear—and why—we step a little closer to understanding our place in the grand dance. Fashion, in this light, is not just surface deep; it is a language alive with connection, spirit, and the endless possibility of expression. And in that, we might find new paths forward, grounded in respect, creativity, and the shared pulse of all life.

Chapter 6: Food and Nutrition

Food beyond Earth is something that stirs curiosity deep inside us. Stories from those who say they've met beings from other worlds open a new window onto how living things might eat, nourish themselves, and even connect with life in strange, beautiful ways. On those distant planets, eating isn't just about filling a stomach. It can be art, it can be spirit, a part of the world's very soul shaped by the planet's quirks and the culture's heart.

Think about this: on Earth, our plates mostly come from animals and plants, the usual suspects. But out there, it's a whole different scene. Imagine foods made of crystal-like forms that sparkle under alien suns. Or plants that glow softly like little stars waiting to be tasted. Even more wild, some beings rely on food that is pure energy, something that feeds the body and spirit at once. These foods are strange to us, but they carry a balance that holds life together on those worlds—nutrients mixed with something we don't yet fully grasp.

The way those beings prepare and eat their meals tells us a lot about who they are. Food is not just something to gulp down in a hurry.

It's sacred, slow, mindful. It can be a ceremony, a meditation. In some stories, cooking is a practice almost like prayer, where every step is honored, every flavor appreciated in silence or soft conversation. Technology is sometimes part of this process, not in a cold mechanical way, but as a tool to bring out the deepest vitality in what they eat. Meals become more than taste—they heal, restore, and even uplift.

Meals often happen in groups, where sharing is more than custom. It's a way to bind hearts and minds. Eating together seals friendship, community, trust. These scenes feel warm and alive, like a dance of voices, colors, and shared hope. Food turns into a bridge between souls. It's a dance where every bite carries meaning, every dish a smile passed around the circle.

Then there's the magic of flavor itself. These alien dishes don't just taste good. They surprise and delight in ways our tongues have never met. Flavors can be sharp and tender, moving and still, like a song that shifts with each note. Some meals look like living paintings, glowing and changing right as they reach your lips. This sensory wonder comes partly from tools that change molecular shapes, crafting food into forms that excite every sense. It's not just nourishment — it's an experience.

The way these beings eat also reflects how they keep themselves strong and balanced. Their food choices respond to more than body needs—they listen to the soul inside. Diets are designed with care for how each life form moves, thinks, and feels. Imagine a meal custom-made for you every day, changing as your energy flows, tuned by technology that reads your state and adjusts like a gentle compass. It's a living art, feeding health and spirit together.

Looking at this vivid picture, it's clear that alien food promises more than strange tastes. It invites us to rethink what eating means here on Earth. Our food is medicine, ritual, and joy all at once, though sometimes hidden beneath hurry and habit. Maybe these stories ask us to slow down and find harmony in how we eat. To honor the bond we share with the earth, with each other, with this fragile life that weaves through every living thing—seen or unseen.

Exploring the foods of other worlds is a reminder that life never fits neatly into our categories. It offers lessons in openness, creativity, and respect. When we listen, we glimpse how food can be more than fuel—it can be a pathway to connection, to healing, to celebration. Whether we find these distant kitchens real or symbolic, their message calls us to hold our meals sacred and live in rhythm with the world.

So the next time you eat, pause a moment. Imagine food as a living presence, part of a grand conversation with the universe. Feel the energy in your hands, the story in your mouth. Food can be a smallest kind of grace, a tiny rebellion against separation. It reminds us all: we belong, deeply, to this living dance, on Earth and beyond.

Chapter 7: Religion and Spiritual Practices

Across the silent stretch of the stars, where earthly eyes can barely reach, stories drift in from beings who walk between worlds. These stories carry a light that pulls us closer to understanding how life—strange and unfamiliar—breathes in the dance of spirit and meaning beyond our home planet. They tell of ways others touch the sacred, ways that ripple out into their daily lives, woven tightly into the very fabric of who they are. Here, spirit is not just a set of rules or ceremonies. It is life itself—a steady pulse that flows through their culture, their thoughts, and their very being.

What strikes you first is how different this spiritual life feels from the often narrow paths we follow down here. There is no single thread, no fixed shape. Instead, these cosmic neighbors seem to open a vast canvas where spiritual expression blooms freely. They encourage us to loosen our grip on what counts as religion, asking us to peek sideways, to imagine a spirituality untethered, alive with possibility and stretching far beyond what we might expect.

One thread common among many is the idea of a consciousness not confined to one being, but swelling out in waves that touch all living things. This is not about who we are as separate selves, but about the great web that binds us all. Beings from those distant worlds speak of a shared awareness—a knowing that the cosmos itself is alive, breathing, and that they are a part of that living whole. They tap into this sense through their minds, reaching out like threads of light to connect not just with each other but with the universal mind itself.

This connection is not cold or distant; it is full of warmth and care. Communication flows silently—telepathically more often than by voice—and it carries not just knowledge but a deep, unspoken love. Imagine a world where learning wisdom is as natural as breathing, where every thought is a gift shared, every feeling an echo sent across the vast cosmic sea.

Their spiritual lives also sing with the music of their worlds. Rituals and ceremonies arise from the soil, the wind, the rhythm of stars circling far overhead. These practices are as varied as the planets themselves. Sometimes they gather in groups, bringing their voices, their bodies, and their instruments together. They create spaces humming with color, light, and sound, where healing and transformation unfold. Patterns of geometry carved in space become sanctuaries, resonating with harmonic waves that spark shifts in consciousness. Movement and art blend into teachings, and the sacred flows not just in words but in dance and song.

Technology, too, plays a role most unlike our own. Far from replacing spirit, advanced tools serve as bridges, helpers for inner journeys and for expanding the fields of perception. Crystals hum with energy; light bends and twists to open doorways unseen by human eyes. These instruments are respected, touched with reverence, and wielded with care. They are gifts shared not for power but for growth—to explore dimensions both inner and outer and to bring the community together in awe and gratitude.

Guides and teachers walk among them, but these leaders do not command or rule. They walk alongside, offering a hand and a mirror, helping others find their own path within this wide and unfolding mystery. There is no single guru or master standing above the rest, but a collective of seekers inspiring one another onward. Education is life itself—a continuous stretch forward into understanding, always reaching toward a fuller picture of existence and the self's place within it.

At the heart of their way is an insistence on love, compassion, and a commitment to service. These are no empty words but living principles that shape every corner of their societies. Service to others is seen as a noble act, one that nourishes both the giver and the receiver. Each individual is encouraged to contribute their unique gifts, knowing that no one thrives alone in this cosmic song. Harmony, cooperation, kindness—these qualities knit their communities together and propel them forward on a shared journey.

Looking out from these stories, we catch a glimpse of a tapestry far larger than our usual frame. They call us to shift our view, to open our hearts and minds to the wider possibilities of spirit. Whether or not we can fully understand these experiences, they remind us of the richness waiting in the vast silence beyond.

Perhaps the greatest gift in these tales is the challenge they pose to us. Not to imitate from afar, but to learn—to turn inward, to notice the rhythm of life around us, to recognize the threads that connect us to all beings. To ask ourselves how we might live with more openness, kindness, and a sense of unity. These glimpses of other lives invite us to weave our own spiritual journey with new care. They remind us that our story, too, is part of the endless unfolding, and that the sacred is never far, always calling us home.

Chapter 8: Love and Relationships

Love among beings from other worlds carries a different tune. It's not just about feelings, or the butterflies we know here on Earth. It runs deeper than that. It's tied to growth, to the way these beings relate not only to each other, but to their communities and the universe. The stories shared by those who have encountered alien civilizations paint love as something alive and vital—woven into the very fabric of existence, shaping the dance of connection in ways unfamiliar to us.

Imagine a love that starts not from words or touch, but from a resonance deep in the soul. A tuning of energies, a meeting of purposes that goes beyond romance or attraction. Many contactees talk about relationships where two beings recognize each other's spiritual frequencies. It's a connection that feels whole and complete, guiding both toward growth and service. Physical forms might differ vastly, but the bond is clear and strong. These beings often share thoughts and feelings directly through telepathy, bypassing distance and misunderstanding. A constant exchange of energy, wisdom, and affection flows freely, nurturing each partner on levels impossible for humans to grasp fully.

Relationships on distant planets morph in shapes and forms unlike ours. They aren't locked into the patterns we know. Partners may come together in groups or pairs, shifting naturally to meet each one's needs. These unions are built on respect and cooperation, not ownership or control. Responsibilities are shared. Resources flow where they are needed most. Imagine communities where love is spread widely, creating webs of support without jealousy or fear. Multiple connections intertwine, forming circles of care that nurture each member. There's freedom in this openness, a freedom that fosters trust and strengthens bonds.

One of the most surprising lessons from these tales is the absence of jealousy. Those painful claws of possessiveness are seen as blocks, stopping hearts from truly opening. Instead, love is practiced as an art of expansion. Each relationship is a unique reflection of a bigger, divine tapestry. People in these cultures are encouraged to look inward, to understand their emotional patterns and release those that don't serve growth. Loving becomes a sacred act, a daily practice of healing and learning. It pushes individuals to evolve and to give back, not just to a partner but to the whole community and the greater cosmic flow.

This cosmic view of love stretches beyond the personal, embracing all life. For many alien civilizations, love is a powerful force that guides the unfolding of consciousness and the balance of existence. It's found in acts of kindness, creativity, and selfless service. Each being contributes their skills and gifts, weaving a collective harmony. The energy of love channels through them, aligning with a universal purpose that sustains and nourishes everything—from the smallest creature to entire planets. The act of loving reaches into this larger dance, moving life forward toward unity and peace.

Harmony and balance keep these societies strong. They value mutual support above all else. Relationships encourage cooperation, with partners working side by side to build lives rich in meaning and respect. Technology and social systems often support open lines of communication and sharing, creating safe spaces where trust takes root. In these environments, love can grow freely, nurtured by security and understanding.

Stories like these invite us to rethink what we believe about love. They show us that connection is far bigger than mere emotion or cultural habit. From the other side of the stars, love emerges as a living, breathing force that links all things—one that beckons us to open our minds and hearts wider. These visions ask us to consider life with a deeper awareness of our place in the wide fabric of existence.

As we look up into the night sky, curious about the mysteries that lie beyond, these glimpses of alien affection offer not just wonder but wisdom. They remind us to honor the ties we have here on Earth, to seek greater kindness and openness in our own relationships. They invite us to see love as a shared journey, woven through the vast, unfolding story of all life.

Maybe we are just beginning to understand what love truly is—not a single thread, but a vibrant, shifting web connecting everything alive. It's a lesson whispered across the cosmos, waiting for us to listen, learn, and open ourselves to its grand and endless embrace.

Chapter 9: Moral Values and Ethics

There is a quiet strength in hearing what other beings—those who walk among stars far beyond our reach—have to say about how they live rightly, how they hold honor in their hearts. Those who have encountered or sensed these visitors speak of ethics not as a strict rulebook but as a living flow, woven into every breath they take and every choice they make. Imagine morality not as a cold list of dos and don'ts but as the very pulse of existence, a compass tuned to the music of growth, balance, and unity. This view stretches our usual thinking and nudges us to see that the universe is vast, not just in space but in the ways life can be wise and good.

What stands out most is how these beings often embrace principles that go beyond personal preference or culture. They speak of a deeper connection, a web that holds all life together. It's a web that asks them to consider how every action touches not just themselves but the whole community and, even wider, the cosmos itself. To live in harmony with this understanding means to carry responsibility—not as a burden but as a natural part of being mature in spirit and aligned with the larger dance of existence.

Picture civilizations where the right choices are reflections of this spiritual balance, where morality is a living mirror showing each being's place and purpose in this great unfolding.

The choices and laws of such worlds show variety, shaped by their unique environments and histories. Yet, a common thread runs through many accounts: an emphasis on qualities like kindness, honesty, and wisdom—virtues that knit individuals together with threads of trust and respect. Technology and social systems often support this way of life, not to control but to help beings better feel and understand what is right. Some advanced societies use what we might call mind-to-mind dialogue and energy sensing to see the weight of their actions before they step forward. Instead of simply reacting to problems or following orders, they pause and reflect on how every deed touches the balance of their homes and community.

What is striking is their gentle way of dealing with mistakes. Rarely do these stories hold punishment as the first answer. Instead, growth and learning take the lead. People are encouraged to see failures as openings, doorways to deeper knowledge and healing. The community becomes a space where forgiveness and support help bring balance back after a misstep. True justice, in this light, flows from understanding and compassion, a process where harmony is restored not by force but by care and respect. When someone makes amends, it is treated as a sacred act, a chance for them to mend their spirit and contribute again to the whole.

The guides and teachers in such places do not stand on pedestals. Instead, wisdom feels like a shared journey, a path each person is invited to walk with care. Individuals cultivate their own sense of right and wrong, ever reaching out to different teachers and experiences to deepen their knowing. Leaders, when present, serve not to command but to help others find their inner light and navigate the complexities of living well. Education is a cherished thread here—ever stretching, always unfolding deeper understanding of life and self.

Love and compassion often light the way as the heart of their ethics. Serving others is not seen as a sacrifice but as a joyful calling. Every person is valued for their unique gifts, which they bring forward in ways that uplift the whole. Acts of kindness ripple outward, spinning a web of connection that holds the community and the cosmos in a gentle embrace. This service is more than duty; it is a sacred role that nourishes both giver and receiver, supporting life's growth and the flow of wisdom throughout all.

Looking beyond our world to these stories, we find an invitation—the chance to rethink how we see right and wrong. The universe, it seems, holds endless ways to live connected, ethical lives, ways rich with love, growth, and harmony. As we turn our eyes upward and outward, these glimpses encourage us to question and reflect on our own moral paths. They remind us to honor how all life threads together, to see justice as more than rules, but as a dance of restoration and care. May these insights inspire us toward a new way of being, one that embraces the fullness of our shared cosmic journey with open hearts and eyes.

In this broad and beautiful dance, ethics become as natural as breathing, flowing through us and around us, calling us to grow beyond what we think we know. Each step forward is a chance to find balance, to weave love and wisdom into the shifting fabric of living. What might it look like if we took these lessons to heart? If we let go of old limits and opened ourselves to a wider harmony, where every choice sprouts from a place of connection and care? The possibilities are as vast as the stars themselves, waiting quietly for us to rise and answer.

Chapter 10: Sports and Recreational Activities

Across the vastness of space, other civilizations move in ways both strange and familiar. When visionaries and those who've touched the stars share their stories, they reveal worlds where sports and play are woven deep into the fabric of life. These are not mere pastimes or simple distractions. Instead, they pulse with meaning, shaping how beings grow, connect, and find joy. What we call leisure is only a shadow of what others understand play to be—a life force that blends body, mind, and spirit.

Think about this: from the stories gathered, it's clear that on other planets, sports often reach beyond physical skill. They are acts of balance, of unity within oneself and with others. Imagine moving your body not just to win, but to tune yourself into the rhythms of nature, the patterns of the stars, the hum of the universe itself. In some places, players manipulate fields of energy, flowing through geometric dances that require sharp focus and deep peace. Others move in sync with sounds and vibrations, creating harmonies that float between members of the group, bringing them close without a word. It's a dance of life that embraces more than muscle—it embraces awareness.

The games and pastimes of these worlds echo their cultures and the land they arise from. They mirror values of making, sharing, and uplifting one another rather than tearing down. Instead of battling to defeat, they create together. Technology doesn't separate; it invites new ways to shape light, color, and sound into playful experiences. Picture a playground where colors shift with your mood or a soundscape that invites you to explore dreams and wonders. These environments captivate and inspire, inviting imagination into a living, breathing experience.

But what really stands out is the way competition fades from view. Winning and losing take a backseat to joy and connection. Sport becomes a sacred act, where showing up and moving together offers deep rewards. The true prize lives in the moment—the skill honed, the bonds formed, the subtle growth felt inside. This presence turns play into a ceremony, an act of honoring the self, the community, and the cosmos. There are no trophies that matter, only the invisible treasures earned through laughter, effort, and shared breath.

On these worlds, sports and play reach further still. They serve as bridges, drawing individuals into a shared pulse with life all around. Communities gather in celebration and creation, building spaces where everyone can participate. Advanced tools and social designs help share what is needed and make room for all voices and abilities. Playing together becomes an expression of unity, a flowing alignment with the deeper currents that move through all living things. It's a way to open pathways where energy can move freely, nourishing growth that touches not just the individual but the whole web of existence.

Alongside the emphasis on harmony and care, these beings prize creativity, discovery, and new ways of seeing. Their games spark minds and hearts alike, turning play into exploration. They tempt participants to push boundaries, to invent, and to weave together fresh ideas. Through sophisticated crafting of materials and tech, experiences arise that feel new and alive, capable of shifting how one understands the universe and one's place inside it. Play here is a teacher, a guide offering insights that ripple through consciousness and culture.

When we pull back to look at this bigger picture, these accounts offer a sunrise beyond our current vision. They challenge our neat categories and limited ideas about fun and sport. They beckon us toward a broader view, one that honors the many ways life finds expression through movement, connection, and spirit. As we reach further into the cosmos and into ourselves, these glimpses remind us to rethink our own games and joys. They nudge us to seek paths where play is whole—a bridge connecting body, mind, and soul, woven tightly with the dance of all life.

This is more than curiosity about other places. It is a call to awaken here, now. To let our pastimes grow roots as deep as our intentions. To find pleasure not in conquest but in discovery and communion. There is sacredness in every breath taken during a game, every smile shared in joyful struggle, every moment where we move as one. The stars above remind us: play is a word far richer than we imagine. It is an invitation to journey—inward and outward—toward the evolving heartbeat of existence itself.

Chapter 11: Music and Art Forms

Imagine for a moment the music and art of beings from far-off worlds. These are not just sounds or pictures meant to pass the time. No, they carry the pulse of existence itself. They speak of emotions, teachings, and connections that reach far beyond what our eyes see or our ears hear. When we listen to stories shared by those who have encountered these extraterrestrial cultures, we're invited to explore creation in a way that stirs something deep inside us—a fresh awareness of what creativity can be.

Take music, for example. It goes beyond rhythm or melody. It becomes a living thread linking the individual to a vast cosmic dance. Many tales describe these sounds as woven with universal balance, quite unlike anything we're used to. It's as if the music itself tunes into the very fabric of the universe, drawing out harmony that echoes through body, mind, and spirit. Think of instruments made from crystal structures, pulsing with energy fields, or even sounds crafted through telepathic waves. These aren't mere tools; they're bridges. Bridges that carry healing vibrations, open doorways to transformation, and awaken a deeper spiritual truth. Each note is carefully placed with awareness of multidimensional frequencies, colors of sound that ripple through space and time.

Arts on these worlds are no less surprising. Visual forms are alive, shifting and responding to who watches them and how they feel. Imagine an image that changes shape or hue depending on your mood or thoughts, drawing you into a space beyond the simple act of looking. The use of light and geometry, color flowing in fractal patterns, brings a sense of awe, perhaps even sacredness. Some creations might even blend organic life itself—living beings woven seamlessly with the art, blurring lines between what is real and what springs from imagination. These artworks expand beyond expectation, inviting the viewer on journeys where perception, emotion, and spirit overlap.

What warms the heart even more is how art and music are rarely solitary experiences in these cultures. They blossom through collaboration. People come together, each offering their gifts not for personal praise but as a shared expression. It's about co-creating something that becomes larger than any single voice or vision. Technology and social harmony provide the tools and structure so ideas can flow freely, weaving diverse perspectives into a collective masterpiece. In coming together to create, individuals find a sacred practice—one that nurtures growth, learning, and connection. The act itself becomes a celebration of community, a path to contributing to something greater—both society and the cosmic whole.

For these civilizations, music and art pulse with deeper meaning beyond personal or communal art. They embody connection with all life. Many describe how music and art serve as vehicles for communicating the unseen mysteries, honoring existence and magnifying shared understanding. Spiritual wisdom flows alongside advanced tech, allowing these expressions to become gateways for energy and insight. When people sing or paint in unison, they aren't just making noise or pictures—they're aligning with a universal flow, a divine current that supports growth, balance, and unity across the cosmos.

Innovation plays a big role too. Instead of settling into fixed styles or themes, art and music remain fluid, always evolving. Creators explore new ideas, pushing beyond current limits, hungry for fresh insights about life and themselves. This constant seeking is supported by materials and tools no less evolving. Through their work, artists spark shifts in consciousness, challenge old ways of thinking, and offer new paths forward. Creativity, then, becomes a journey—not a destination—inviting survival and transformation, always opening new doors and revealing new horizons.

When we look at these stories as a whole, they invite reflection. They urge us to loosen the boundaries of our own imagination and think beyond Earth's familiar patterns. This perspective honors a living diversity that stretches across the universe. It encourages us to find in our own music and art a deeper integration of heart, mind, and soul—an expression connected with life itself and the vast cosmic flow. Creativity, in this view, is not just a human gift. It's a universal pulse, a sacred rite woven through all beings, across all worlds.

So, let these insights warm your spirit and spark your curiosity. Let them remind us that in the endless story of existence, music and art are not just distractions or hobbies. They are threads of unity and awakening, calling us to listen more closely and create more openly. They invite us to walk with grace on a path that honors our links to each other and to the cosmos beyond. In their wake, we find more than beauty or wonder—we find the possibility of harmony, connection, and a deeper meaning waiting to unfold.

Chapter 12: Technological Development

When you gaze up at the night sky and wonder who else might be out there, your mind naturally drifts to what their lives are like and what kind of tools they use. The stories told by people who say they have met beings from other worlds offer a view that's quite unexpected. The devices and technologies are much more than simple machines. They hold a place deep inside the hearts, minds, and very souls of those beings — shaping their culture, their spirit, and their journey across the vastness that stretches beyond.

What stands out is how these technologies act in partnership with the natural world. They are not just gadgets to grab resources or control an environment. Instead, they grow out of the land itself, taking part in its balance. Imagine tools made from elements that blend into the earth, energy drawn from forces that never run out and cause no harm. These civilizations tap into sources of power that we barely glimpse — glowing crystals, fields of vibrating energy, even living creatures become part of the tools.

The technology feels alive, moving with the pulse of the universe. Imagine engines that turn solar storms and cosmic rays into energy that never fades, never leaves scars. This is technology living with nature, not in battle against it.

For these beings, science is a different kind of story. It's not about cold charts or hard proof alone. It is about finding harmony, a balance in well-being — for the individual and all who share space together. They use knowledge to raise consciousness, heal beyond the body, reaching into mind and spirit. Healing comes through light, through sound, through patterns that vibrate in tune with life itself. It's a whole approach to health that sees us not as broken or alone but as threads woven into the great flow of being. Every tool, every practice moves to support this balance — from health to how they talk and change, all woven through technology that nurtures life.

And the way they create new ideas is striking. It's no secret lab, no lone genius hidden away. Instead, it's a shared dance. Skills are passed freely. Ideas bounce from one to the next like a living river. Tools rise from this collective pulse, made for everyone to use and grow with. The act of creating holds a sacred rhythm, almost a ritual that honors the bonds between people. Technology becomes more than an object—it becomes a space where connections spark, where minds meet and lift each other higher. Their societies foster this open flow, letting many voices and dreams mingle and reshape possibility. It feels like science with a heart — driven by more than just fixing problems or making things work.

This sense of connection runs even deeper. Technology is not something apart from life but is life itself. Their stories show invention tied to celebration, to spirit. The tools and knowledge become a way to explore the vast mystery of existence and to share that joy with others. Resources, wisdom, experience all move freely, like gifts in a network built on respect and generosity. To invent is to step into a dance with something larger, a movement where energy stays balanced with the wider cosmic beat.

To explore is not merely a goal but a way of living for these beings. They move forward with curiosity, always searching, always testing, never settling. Their technologies are alive, changing right alongside them. This living process calls for new eyes on the world — for pushing ideas beyond what they once thought was possible. It is an invitation to open wider, to expand not just machines but our sense of reality itself. The technology flows, shifts, and grows, just like life — never fixed, never caught in one place.

These stories give us fresh insight into what technology might really be. They ask us to imagine something gentle, something that thrives on working together, something beautiful — not just a race or a game of power and control. They encourage a vision of the future where our creations feed the soul, the community, and the earth, not just meet immediate needs. When we look up at those stars, these visions softly ask us: what kind of tools do we want? What kind of world can we build if respect and balance guide our hands?

The vastness above stretches out, packed with secrets and endless promise. The tales told at the edges of our own experience are more than fantasy. They carry whispers, gentle lessons on another way to live—as here, now. They invite us to tune into a larger rhythm, to bridge the space between science, spirit, and the living world around us. In taking in these stories, we find no clear answers. Instead, we find new questions — questions that light the way forward. A path that honors the living web we all share, where technology lifts us all toward a greater destiny among the stars.

Chapter 13: Innovation and Scientific Achievements

Across the far reaches of space, beyond the familiar blue of our planet, there are stories—deep, vivid tales told by those who claim contact with beings not of Earth. These narratives speak of civilizations whose discoveries and technologies are unlike anything seen here. Yet, what stands out is not just the gadgets or inventions themselves, but the way such progress seems to flow from a living sense of purpose, one that stitches together culture, spirit, and the unfolding of existence itself.

Imagine technologies born not from a cold, mechanical urge to conquer nature, but from a profound desire to move in harmony with the energies and rules that pulse through the universe. These extraterrestrial achievements often reflect a kind of science that moves fluidly with unseen currents—energies of consciousness, physics beyond what our textbooks cover, and dimensions layered over the familiar three. They seem to play with forces that blend healing and knowing in ways that ripple through whole communities, perhaps even across star systems. There are whispers of ships that ride cosmic flows, carried by energy fields no rocket fuel could touch, moving effortlessly across great stretches where light itself bends and weaves.

The diversity of these scientific paths is staggering. Each culture crafts its own approach according to the land it calls home and the values it lives by. But an echo runs through many tales: a commitment to balance, connection, and the growth of both the individual and the collective. The science here does not stand apart from life—it breathes with it. Some civilizations use beams of light, sounds, and subtle vibrations as tools of healing. Not just for the body, but for the soul and mind, restoring rhythm and harmony by tuning into the universal pulse that connects all beings.

What draws my attention most is the spirit of togetherness in these accounts. Science is not a silent pursuit locked in a lab or carried out by lone geniuses. Instead, it unfolds as a shared journey, each person bringing their piece of insight, their unique skill, weaving these into greater understandings. Technologies and social systems seem designed to foster open sharing, to merge perspectives and spark creativity that lifts everyone. Innovation is a sacred act, a dance of learning and giving, a deep offering to life itself. The pursuit of knowledge is village work, cosmic work, unified in service to growth and healing.

Further still, the role of scientific progress extends beyond discovery. It acts as a bridge between beings, a way to honor and celebrate the mysteries spread across the stars. Spiritual practices and communal structures support this flow, helping to share wisdom and resources freely. The process of uncovering new truths feels like stepping alongside the greater flow of existence. Each breakthrough becomes a gesture of alignment—not just with abstract laws, but with the divine pulse that carries all living things forward.

These visions also portray science as a living, moving stream, never static or closed. Explorers from distant worlds keep pushing boundaries, not simply to know more, but to deepen connection with all that exists. Their methods might employ materials and ideas far from our grasp, yet the intention behind them is clear: to awaken, to open new doors within consciousness, and to pioneer fresh ways of being that reflect the vastness of the universe itself.

So when we consider what innovation and discovery might look like beyond Earth, these stories invite us to rethink. They ask us to see science not as a lonely climb toward power or control, but as a collective dance that respects the ties binding us to each other and to creation at large. They remind us there are countless ways knowledge can grow and transform, nurturing life in all its wondrous forms.

As we gaze outward and step deeper into cosmic mysteries, these insights offer a beacon. They challenge us to approach our own quests for understanding with openness, humility, and a greater sense of unity. Maybe through that shift, our inventions and ideas will become pathways to harmony, honoring the sacred journey we all share across time and space.

In this light, the universe stops being an endless void of unknowns and becomes a living network of potential—a cosmic neighborhood where every discovery is a step closer to the whole, every innovation a thread in a vast, unfolding story. It is a vision of science that sings with life, spirit, and the quiet power of hearts joined in curiosity and love. And that is a story worth listening to, wherever we find ourselves in the grand dance of existence.

Chapter 14: Environmental Interactions and Habitability

When we listen to the stories of those who have met beings from beyond our world, a new vision takes shape—one that feels alive and close, not distant or strange. These accounts reveal ways of living that flow with nature rather than pushing against it. On some far-off planets, caring for land, water, and sky is not just a task to get through but a powerful way of being. Their care spreads through beliefs, daily habits, and their ties to the vast universe around them.

Many travelers who share these experiences talk about balance as the foundation of these societies. The beings they met know how to live alongside their world, using tools designed to give back instead of just take. Their materials often heal and grow on their own. Energy comes from endless sources, and their homes are alive in form and spirit, shifting and breathing like plants or coral reefs. Picture buildings that feel alive—not rigid shapes made of steel and stone but spaces that grow, change, and tune in to the earth beneath them. In these places, stone, wood, and living space blend as one, each supporting all the others.

Life in such places isn't just about surviving harsh storms or strange skies. It is about weaving soul and craft into the environment so it welcomes and holds every being. Heat that burns where we live is embraced elsewhere; gravity that tugs oddly is met with something fluid and whole. Their worlds are as different from each other as a mountain is from the ocean. When hardships arise, they turn to art, science, and an ancient knowing of nature's flows. Their farming looks nothing like our straight rows in fields. Instead, they nurture growth through light, energy, and crystal patterns that strengthen life without tearing soil apart. They are partners with the earth, loyal to cycles and rhythms that might feel foreign to us but are as steady for them as breathing.

What shines brightest in the tales is how caring for the environment belongs to the whole community. It is never left to a few. Everyone joins hands to protect the land, waters, and creatures around them. Knowledge moves freely. Ideas rise and spread like wind. Cooperation here is more than working together—it carries a sacred weight. To tend the land is to care for the heart of their people and, through that, the pulse of all things. What one does touches countless others, seen or unseen. This is a net of respect and responsibility, a dance of care where every step matters.

These stories remind us the environment is no mere backdrop. It lives and breathes at the center of culture and connection. Songs, dances, and discoveries emerge from this link. Deep wisdom and fresh ideas unite in ceremonies that weave energy and spirit among all beings and their surroundings. For these beings, caring for the earth wakes the mind and opens the heart. Nature sings its old songs, and those who listen move with the music, inviting new insight and life into every moment. Here, spirit and ecology rise as one, bright and unbroken.

Alongside this harmony exists a restless openness to change. Traditions do not fossilize but grow and shift as new cosmic truths unfold. Curiosity flows through their care for the planet like a river. They question old stories, cross borders, and create ways of living that honor both their world and their inner knowing. Their tools and methods sometimes seem like magic; other times, they feel as ancient as stone and sky. Every act of tending the earth invites seeing deeply, thinking clearly, and being fully alive.

Taken all together, these visions of life beyond Earth offer us more than new knowledge—they hold a gentle teaching. They show a way to care for the world that is joyful and shared, born from respect and belonging, not struggle or use. These tales invite us to imagine places where land and life grow in step, where caring for nature opens the door to understanding the universe's heart. When we look to the stars, they ask us to listen closely, honor all connections, and build homes that hold the rhythm of the planet and beyond. In this, may we find a path to walk softly, live well, and join the endless dance of life threading through all worlds.

Chapter 15: Habitability and Living Conditions

When we listen to those who talk about life beyond our world, we find stories that paint a very different picture of what it means to live somewhere. It is not just about having a roof over one's head. On many distant planets, the idea of home is tied deeply to how life flows, how spirit dances, and how worlds sing their own unique songs of balance and belonging. These tales invite us to step outside our usual way of thinking about space and structure and imagine the endless ways life could shape itself in the great expanse.

One theme that keeps coming up is this idea of blending in with the world itself. Imagine homes growing out of the land like plants do, shaped not by hammer and nail but by living beings and crystals that stretch and change with time. These places are not cold constructions; they breathe and adjust, made from strange materials that are alive in a sense. Energy moves through them, supporting those inside in a way that feels natural, not forced.

There is a sense of partnership — the building and the planet cooperate to create something that lasts and thrives.

The conditions life adapts to can be wildly different. Some planets are scorchingly hot, others shiver with cold no human could bear. Some have thick, heavy air; others are thin and strange. Gravity might pull gently or press down like the weight of a great story on your chest. Radiation might dance like sunlight or bite like a storm. Despite this, these beings manage to carve homes that protect and care for them. They use clever tools and practices that go beyond the physical, sometimes spiritual, to shape spaces where health and happiness are front and center. Fields of energy, invisible to our eyes, can hold temperature steady or clear away harm, crafting comfort in worlds that seem hostile at first glance.

What stands out isn't just the buildings themselves, but how these communities come together to create and maintain their homes. There is a spirit of sharing woven through these stories, where knowledge, resources, and skills flow freely among neighbors and friends. Living spaces are more than private refuges; they are fruits of collective care and attention. People gather their talents to build not just for themselves but for the whole. Architecture becomes a way to learn about each other and shape the bonds that tie souls together. And the act of creating a home is never ordinary — it is a sacred dance, a chance to tap into the greater wisdom that pulses through life.

In these societies, buildings embody more than shelter. They express connection — a web that includes everyone and everything, from the smallest life form to the farthest star. Crafting a living space is a way to honor this vast unity. Homes stand as symbols, not just of personal safety but of shared journey and celebration. Advanced views and practices support this vision, tools and rituals that help people share what they have, grow in wisdom, and design places that uplift and inspire. Building is communion — a holding open of heart and mind to the flow of universal energy that guides and sustains all living.

Another strand that shines through these stories is the hunger to explore. Homes are not fixed or final. They change, grow, and evolve alongside those who live in them. The act of creation is not static but a continuous reaching outward. New materials, new energies, and new ideas are brought into the fold. The architectures of these worlds reflect an ever-widening understanding of existence — awakening new ways of seeing and being. Building is more than shaping space; it is stretching one's mind and spirit beyond old borders, opening to fresh insight and possibility.

Looking at all this, we are invited to rethink what it means to live well—not only in far-off places but here on Earth. These stories are reminders that our own houses and cities could carry more of this richness and care. They ask us to see buildings and communities not just as objects or containers but as living, breathing parts of a much bigger story. If we listen, we might find ways to bring harmony, beauty, and deeper respect for life into the spaces we call home. They call us to remember that everything is connected: earth, sky, water, and spirit.

So as we gaze upward and ponder what life beyond might be like, we are also looking inward. The lessons whispered by other worlds encourage a gentler touch, a more fluid dance with the energies around us. They speak quietly but firmly about the possibility of homes that nurture all living things, that grow with us as we learn to weave together the many threads of existence. In this lies an invitation — to honor the sacred in every stone and star, and to create places on our own planet that reflect the vast, unfolding miracle of life itself.

Chapter 16: Consciousness and Spiritual Beliefs

There's a fascinating story unfolding beyond the stars, whispered by those who say they've made contact with beings not from this world. These stories paint a picture of consciousness and spiritual life far removed from what we usually imagine. They speak of minds and hearts wired differently, worlds where inner life isn't just personal—it's the thread running through the fabric of their culture, their growth, their very way of being. This isn't about strange aliens reciting ancient texts or following rigid rules. It's about living awareness that spreads out, connects, and shapes everything around them.

What stands out most is how these beings seem to live in a state of deep connection. They don't see themselves as isolated dots, separate from each other and the world. Instead, they sense themselves as parts of a vast, living whole. Their awareness stretches beyond their individual self, touching the life flowing through the stars and the spaces between. It's not just an idea but a felt experience—a knowing that we are all linked, part of one great unfolding story. This feeling guides their beliefs and actions, drawing them toward harmony with the universe and with one another.

Imagine civilizations that speak not just with words but through pulses of thought and feeling. Telepathy isn't a rare trick; it's how they share their minds and hearts. Conversations bridge vast cosmic distances without barriers. They tap into a universal mind, a wider source that carries wisdom and kindness across galaxies. This constant sharing weaves communities not by laws or borders but by understanding and love. What's shared grows stronger, not weaker—it's as if their very lifeblood flows through this exchange, feeding creativity and compassion.

Their spiritual paths bloom according to the special conditions of their worlds. Each society mirrors its home in the values it holds dear. Some revolve around balance and respect, others honor truth and kindness. They develop ways to move beyond the physical, exploring dimensions that ripple alongside our own reality. Through focused practice, they learn to shape energies — light, sound, vibration — to heal, to awaken, and to open doors to new awareness. These methods aren't mystical secrets locked away. They're tools for anyone who wants to connect deeper, to touch the sacred threads that bind all life.

In many accounts, spirituality is not a solitary quest. It's something people do together. They come with their strengths and insights, weaving diverse experiences into a shared journey. This collective search deepens their growth and expands their connection. It's a dance of minds and spirits, lifted by social structures and technology that encourage sharing and learning side by side. Spiritual practice becomes a space where differences are celebrated, where ideas spark new insight, and where transformation happens in the company of others. This teamwork in spirit reflects an understanding that no one grows alone — our joy and struggles ripple outward, shaping the community just as it shapes us.

Life on these worlds pulses with a wider view. Consciousness and belief don't stop at the edge of the self. They spread through neighborhoods, cities, and societies, coloring every gathering and feast, every conversation under alien skies. Connection and celebration go hand in hand. Spiritual life builds bridges, opening hearts and minds to the unknown with trust and reverence. These traditions foster generosity and wisdom, cultivating peace and unity. Engagement becomes a sacred rhythm that brings the community into alignment with a greater harmony flowing through all things.

Beyond connection, there's an eager embrace of discovery. Spirituality here isn't fixed but fluid, always growing and changing. The journey itself matters more than resting points or answers set in stone. Individuals stretch their inner horizons steadily, seeking new ways to know the universe and their place within it. They apply advanced techniques, blending ancient insight with emerging methods to create fresh invitations for growth and change. These experiences push boundaries — shifting beliefs, expanding minds, inspiring new forms of being.

This ever-evolving path is not separate from daily life but woven into the pulse of civilization. Questioning, seeking, opening — all form the pillars of this approach, urging beings onward in their quest for meaning and wholeness. Spiritual exploration becomes a way to challenge old patterns, light new fires, and invite the unexpected into the dance of becoming.

So what do these glimpses tell us? They invite a pause, a breath of wonder. They remind us our familiar definitions of spirit and consciousness might be small windows to a far bigger view. Beyond Earth, life is rich with possibility — different expressions of connection, purpose, and growth all intermingling under the vast sky.

Chapter 17: Existential Purpose

The extraterrestrial stories ask us to stretch our vision, to honor diversity in how life can awaken and find meaning. They nudge us toward a spirituality grounded in togetherness, openness, and respect for the countless ways consciousness unfolds in the cosmos. They encourage us to reflect on our own paths, to live with heart, and to seek harmony not only with one another but with the many layers of existence embracing us all. Through this wider view, the sacredness of life becomes clearer, and the journey itself shines as a gift shared by all who dare to seek beyond the stars.

Understanding how otherworldly beings view the meaning of their existence invites us on a fascinating journey. It shows that meaning is not just a topic for thinking or debate but something alive in the heart of their cultures, spirits, and the very unfolding of their galaxies. These stories and messages shared by many who claim contact with those from afar reflect a world full of possibilities for how purpose can be lived and known.

One of the most striking ideas is how these civilizations emphasize a deep harmony with the universe itself. They talk about a sense of being part of a larger design. This is not just about what one here or there wants, but about being connected to all life and moving toward the greater well-being of many. This kind of meaning stems from a feeling inside that guides actions to support community, evolution, and balance. It's as if their souls know they are threads in a vast living whole. From these accounts, beings who have grown in spiritual wisdom carry a responsibility—not just for themselves but for all life. Their life's purpose mirrors their inner state, a sign of how far they have come in understanding their place in this cosmic unfolding.

What is also beautiful is how varied these purposes can be. Each world, shaped by its own skies, landscapes, and lessons, carries its own ideas about what matters most. In many stories, values like kindness, honesty, and deeper knowing are not just moral ideals but daily practices. People are encouraged to live with these traits as cornerstones of their journey. Sometimes this work means exploring dimensions beyond the physical, playing with energy, or learning ways to heal and transform. Through these acts, they become not just travelers but creators of sacred space and growth, steering their souls and societies toward new heights of consciousness and understanding.

Something else stands out—the way purpose is often shared. Life there feels less like a solo climb and more like a gathering of many hands and hearts. It seems their worlds have found ways to weave individual passions and gifts into a chorus of meaning, strengthening each other as they go. This kind of pursuit, treated almost like a sacred ritual, opens pathways for learning and evolving together. It is through walking this path as a community that they feed not just their own light but the light of all life surrounding them. Social structures and advanced tools support this sharing, making space for ideas to flow and blossom in unimagined ways.

Beyond personal fulfillment, these teachings show that purpose lives in connection. It breathes in the rhythm of communities where everyone's quest adds to the collective harmony. This shared journey serves as a bridge linking individual growth to the larger dance of existence. It's also clear that such societies nurture this bond through sophisticated spiritual and social practices that encourage generosity, wisdom, and transformation. Embarking on one's path here is understood as a sacred alignment with something greater, a force that moves energy, insight, and life forward in waves of balance and joy.

Exploration and growth form another key thread. In these visions, purpose is never fixed or static. It flows and changes, inviting beings to keep searching, learning, and expanding what they know about themselves and the universe around them. The technologies and practices they have developed help create moments that shift perspectives and open new doors in the mind and heart. They see living purpose as a journey of consciousness itself, challenging old limits and sparking fresh ways of seeing and being. This ongoing quest keeps them vibrant and awake, a dance of curiosity and wonder that honors the sacred work of discovery.

These stories offer more than just ideas about distant lives. They invite us to open our minds and hearts to what could be beyond our familiar walls. By sharing these different ways of understanding meaning and purpose, they ask us to rethink our own paths. They remind us that life might be far richer and more connected than we often realize. It stirs a longing inside us for a deeper harmony, an integration of self and other, a sacred respect for all threads woven into the fabric of life.

Chapter 18: Interplanetary Roles and Relationships

The realities of other worlds encourage us to reflect on our own place in this vast mosaic. The universe, with all its wonders, hints at endless stories waiting to be told and lived, stories that might help us understand ourselves a little better and grasp the unfolding journey that includes us all. The search for purpose among other beings reminds us of our shared heartbeat with the cosmos. It lights a path to a more open, connected, and meaningful life—not just for us but for every form of living, every spark of consciousness reaching toward the light. This is a search that calls for courage, for openness, and for a humble willingness to learn from others, near and far. It asks us to see beyond ourselves and honor the sacred dance of all life moving together through time, space, and the mystery that frames it all.

Looking out beyond our world, the stories told by visionaries and those who've touched the edges of space open up ideas that make us think differently. Extraterrestrial civilizations, as described by these voices, don't just trade or talk like diplomats. They weave relationships that root deep within their culture, spirit, and a collective journey of growth that spans the stars. This challenges how we view relations between nations on Earth, inviting us to imagine a cosmos filled with potential to unite, share, and grow together.

One of the most striking patterns in these accounts is the idea of togetherness. Across planets, beings come together not as separate pieces but part of a whole, exchanging knowledge and lifting each other upward. This sense of connection isn't a side note; it shapes their way of life. It's about responsibility—caring for the growth and balance of all living things. Their roles on this cosmic stage reflect a maturity, a deep alignment with the forces that bind universe and soul.

Each civilization brings something unique to this dance. Imagine worlds shaped by their landscapes, histories, and inventions. Their relationships mirror these differences, founded on respect and a desire to move forward as one. They don't compete for power but contribute what they're gifted with—technology, wisdom, and spiritual insights—to help the collective move toward a richer existence. Sometimes, they form groups like councils or alliances, gathering to face common puzzles, explore vast mysteries, and nurture life in all its forms.

Collaboration pulses at the heart of their interactions. This is not a simple handshake; it's a sharing of life's richness in every form. From knowledge to resources, beings join in pursuit of dreams that reach beyond their own needs. Such cooperation is supported by social systems and communication methods that bridge worlds and ideas, making it easier to bring diverse voices into harmony. This way of working together is sacred, a practice where growth and healing flow both ways, mingling to restore cosmic balance.

But there is something deeper still. In these stories, interplanetary connection is more than cooperation. It's a celebration, a way to communicate soul to soul and to unlock the unknown. These connections embrace wisdom and offer experiences that inspire change and wonder. By participating in these relationships, civilizations align with a greater will, tuning into the flow of life's energy that supports all beings. It's a dance that knits together the living cosmos, unfolding new chapters in the story of existence.

The drive to explore and discover is another recurring theme. Interplanetary connections are never static. They shift, grow, and evolve as beings push boundaries, seeking to understand more about the universe and their own place within it. Advanced tools and practices let them forge new paths and enter realms of experience that spark transformation. Here, cooperation between worlds acts as a gateway to expanding consciousness, challenging long-held beliefs, and inspiring fresh ways of living and thinking.

When we gather these threads, a vivid picture emerges. Beyond Earth lies a vast, rich spectrum of life, a powerful counterpoint to narrow views focused only on competition or survival. These stories remind us of the immense variety possible when beings choose to work together. They encourage us to see our roles from a broader vantage point, urging us to foster a cosmic cooperation rooted in respect and understanding. In doing so, we honor the deep connections that hold life in balance and recognize the sacred nature of this universal journey.

As we continue to gaze outward and explore unknown horizons, these insights serve as beacons. They show that the future can hold deeper unity and shared purpose among stars, inspiring us to craft a path that values every voice, every culture, and every spark of life. This is not just a dream of distant worlds but a call to reflect on our own place in a universe that beckons us to connect, grow, and live in harmony.

Chapter 19: Challenges and Conflicts

When we look beyond our world to other civilizations scattered across the stars, a surprising truth emerges. The problems and struggles they face aren't mere setbacks or fights. Instead, they become openings for growth, for learning in ways that stretch beyond the mind, touching the spirit. These stories from visionaries and those who've had contact with these beings invite us to rethink what challenges mean. Across countless worlds, difficulties might seem to threaten existence, yet they are part of a grander process—a chance to evolve together, to rise and reshape.

Imagine a place where the environment shifts suddenly or resources grow scarce. Instead of panic or despair, they meet these changes with a calm curiosity and a steadfast spirit. They gather as communities, weaving their strengths together, leaning on one another. This isn't just about survival. It's about stepping into a larger story, where every test shapes their collective soul. They have means—both spiritual and technological—that help them adapt, sometimes in ways we might barely understand. What's remarkable is their willingness to see problems not as punishments or curses, but as invitations. Invitations that open new doors to understanding nature, themselves, and the ties that bind them.

Conflicts do arise. After all, life is full of differences—from ideas to needs, to emotions. But here, too, the way these beings handle disagreements sheds light. Theirs is not a simple arguing match or a struggle for power. Instead, they dive into dialogue—real talk carried on with openness and empathy. Technologies that carry thoughts straight into minds, energies that calm and clarify, tools that turn friction into harmony. It's a dance of respect and connection, a conscious effort to heal rifts and build stronger bonds. Conflict becomes a springboard, pushing them into clearer understanding and deeper compassion. Imagine that: fighting less to win, and more to grow together.

One can't help but notice how much emphasis is placed on working as one. These stories tell of societies where problems are not carried alone but shared. A challenge becomes a gathering point, a chance for people to bring what they can offer—ideas, help, insight—and stitch a bigger picture. The act of solving problems together becomes almost sacred. These moments forge bonds, remind everyone that the wellbeing of the individual is inseparable from the wellbeing of the whole community—and, even further, the cosmos itself. In such cultures, no one walks that path solo. Every hand extended, every voice raised, lifts the entire group upward. The very act of facing struggles together fuels evolution, awakening growth that's both personal and collective.

Yet these challenges stretch beyond mere community survival or local concerns. For some of these civilizations, overcoming difficulty ties them to something vast and meaningful. They see adversity as a thread that draws all life together. When they confront hardship, they are also celebrating resilience—they share their stories, passing wisdom like a living fire. This process becomes a way to connect deeply, not only with each other but with the universe's mysteries. Spiritual practices and social patterns support this, creating rich spaces where lessons and resources flow freely. Facing conflict is not just handling a problem; it's aligning with the larger cosmic dance, letting energy and wisdom flow in ways that nurture growth and balance.

Exploration and discovery fuel their response to challenges, too. These civilizations often view obstacles not as dead ends but as ever-changing puzzles to unlock. Each difficulty brings a fresh chance to see new truths—to redefine what's possible. Technologies and spiritual tools blend in harmony, letting them craft solutions that surprise and uplift. Engaging with problems becomes an act of expanding consciousness, breaking through old views and opening minds to fresh perspectives. This never-ending quest teaches them to think differently, to be alive to new ways of existing.

So many stories from beyond our world call us to rethink adversity here on Earth. They remind us that problems are not just burdens to carry but moments that shape who we are, both individually and as communities. These glimpses encourage us to move toward a more connected way of living—with each other, and with all life. The struggles we face can become a path toward unity and deeper understanding, if we learn to embrace them as part of something much bigger.

As we keep looking outward, reaching toward the stars, these tales speak clearly and simply. They reveal that life, in all its forms, holds endless lessons in the challenges it sets before us. And through these lessons, we can glimpse a future where harmony is not a dream but a practice—a life shared fully with one another and the vast cosmos we belong to.

Chapter 20: The Future of Interplanetary Life

Imagine a future where life stretches far beyond the boundaries of Earth, reaching out to other planets, stars, and even dimensions we barely grasp today. This isn't just about traveling through space, though that's part of it. It's about a deep shift in how beings understand themselves and each other. Not just machines and technology moving forward, but consciousness itself growing and weaving new ways for life to exist and relate on a cosmic scale. Think of it as a great unfolding—not some fixed outcome shaped by the patterns we know, but a living, breathing story where each new page brings surprises and changes the whole narrative.

Many who have reached beyond ordinary experience—those visionaries and contactees—paint pictures of a universe vibrant with connection. They describe a future where beings born of different stars and even non-material realms come together, not as settlers or invaders, but as members of a community built on shared respect, love, and a common understanding of purpose. It's less about borders and more about bridges. Telepathic communication is often mentioned—not merely between individuals but as a way to touch a collective mind or heart that pulses with cosmic wisdom. Conversations happen beyond words, flowing like waves across the vast sea of shared awareness. This harmony creates spaces where knowledge, energy, and kindness flow effortlessly, supporting growth far beyond what we've seen before.

Technology, too, becomes much more than tools or machines. It evolves to work hand in hand with the environment of each planet, blending naturally with the life there rather than dominating it. These inventions draw from sustainable energies and often tap into layers of reality that go beyond the physical. Imagine spacecraft propelled not by fuel but by the energy fields woven through space itself, allowing fast and gentle travel across immense distances. Healing technologies that help beings restore balance and vitality come into play, as do devices that open windows into other dimensions or realities. Every tool reflects the values of the civilization that created it.

What stands out in many stories is the spirit of collaboration rather than competition. Different worlds, diverse species, and multiple dimensions share what they know and pool their resources for common goals. There's a sacred quality to this cooperation, an understanding that each participant offers unique gifts that enrich the whole. When beings work together, it's not simply practical; it's a form of communion. Social systems and communication networks make it easier to weave together these varied perspectives and find new solutions that help all, not just a few. This attitude shifts what progress looks like. Instead of one culture imposing its will, every voice is heard, and every contribution matters.

Beyond this, the idea of unity stretches even further. Interplanetary life isn't seen as isolated interactions or something to conquer. It blossoms into a cosmic dance of relationship, a celebration of existence itself. Spiritual practices support this, guiding communities to share not just goods and knowledge, but also experiences that transform and uplift. At the center of these traditions is the awareness that cooperation links beings to the larger divine flow—the life force that runs through the entire universe. This flow brings energy, insight, and opportunities for growth that lift all life toward balance and harmony.

Exploration remains alive and well. It's not just about mapping terrain or visiting new places, but a journey inward as much as outward. Every contact offers a chance to expand the mind and heart, to look at oneself through fresh eyes, and to discover new ways of being. With advanced practices and technologies supporting these quests, every step forward opens doors to fresh perspectives. Curiosity becomes a cosmic pulse, driving civilizations not merely to survive but to evolve continuously through their interstellar relationships.

As we consider these future possibilities, it's clear the universe holds more than we can easily imagine. It reminds us that growth happens in many dimensions—of thought, spirit, and community. Perhaps most importantly, these visions call us to remember our own role in the larger story. This invitation asks us to think about the future not as a distant destination but as a path we shape together. A path where respect, connection, and love guide us as much as invention and discovery.

The future of life beyond Earth promises to be as much about opening the heart and mind as it is about crossing distances. It asks us to grow into beings capable of remembering unity, even when scattered across the stars. It asks us to honor the sacredness in every form of life and in the dance of cosmic change itself. In the end, these visions offer not just hope but a call—to co-create a future woven from shared light and wisdom, and to celebrate the unfolding mystery that surrounds us all.

Glossary

Adaptation: The process by which beings adjust to their environment, often involving biological, cultural, and technological evolution to thrive in diverse and challenging conditions.

Advanced Technologies: Innovative tools and systems that harmonize with natural environments, often incorporating sustainable energy sources and multidimensional realities to support life and consciousness.

Alien Civilizations: Societies originating from planets other than Earth, characterized by unique biological, social, and spiritual adaptations to their environments.

Art and Music: Creative expressions that transcend traditional forms, often incorporating telepathic communication, energy fields, and multidimensional experiences to foster connection and spiritual growth.

Balance: A state of harmony and equilibrium within oneself, society, and the environment, often pursued through spiritual practices, technology, and communal living.

Biological Diversity: The variety of life forms and biological adaptations found across different planets, often shaped by unique environmental conditions and evolutionary paths.

Collective Consciousness: A shared awareness or mental connection among beings, allowing for telepathic communication and a sense of unity and interconnectedness.

Community: A group of beings living cooperatively, often characterized by shared resources, mutual support, and a deep sense of connection and purpose.

Consciousness: The state of awareness and spiritual growth, often explored through meditation, telepathy, and other advanced practices to achieve higher understanding and unity.

Cooperation: The practice of working together harmoniously, emphasizing mutual support, shared goals, and the well-being of the collective over individual gain.

Cosmic Dance: A metaphorical expression representing the interconnected and harmonious movement of life and energy across the universe.

Cultural Systems: The diverse ways of life, traditions, and practices that define societies on different planets, often reflecting their unique environments and spiritual beliefs.

Energy Fields: Invisible forces or vibrations that permeate the universe, often harnessed for healing, communication, and technological advancements.

Environmental Interactions: The relationship between beings and their natural surroundings, emphasizing sustainability, respect, and harmony with the planet's ecosystems.

Exploration: The act of seeking new knowledge, experiences, and connections, both within the physical universe and the inner realms of consciousness.

Extraterrestrial Life: Life originating from planets other than Earth, encompassing a wide range of biological forms, societies, and spiritual practices adapted to their unique environments.

Harmony: A balanced and peaceful state of existence, often achieved through spiritual practices, communal living, and respect for all life forms.

Healing: The process of restoring balance and well-being, often involving advanced technologies, spiritual practices, and communal support.

Innovation: The development of new ideas, technologies, and practices that enhance life, consciousness, and societal progress.

Interconnectedness: The state of being connected on a deep level, often through telepathy, shared energy fields, and a sense of unity with all life.

Interplanetary Relationships: Connections and interactions between different planetary civilizations, characterized by cooperation, shared knowledge, and mutual respect.

Moral Values: Principles and ethics that guide behavior and societal norms, often emphasizing kindness, honesty, and the well-being of the community.

Multidimensional Realities: The existence of multiple layers or dimensions of reality, often explored through advanced spiritual and technological practices.

Purpose: The sense of meaning and direction in life, often tied to spiritual growth, communal well-being, and the pursuit of harmony and balance.

Spiritual Practices: Activities and rituals aimed at fostering spiritual growth, connection, and harmony, often involving meditation, telepathy, and energy work.

Sustainability: The practice of living in harmony with the environment, ensuring that resources are used responsibly and that the planet's ecosystems are respected and preserved.

Telepathy: The ability to communicate thoughts, feelings, and ideas directly from one mind to another, often used for deep connection and shared understanding.

Unity: A state of oneness and interconnectedness, emphasizing the collective well-being and the harmony of all life forms.

Book Review Request

Dear reader,

Thank you for purchasing this book! I would love to know your opinion. Writing a book review helps in understanding the readers and also impacts other readers' purchasing decisions. Your opinion matters. Please write a book review!

Your kindness is greatly appreciated!

Also Written by the Author

1. Alien Worlds: Biology, Society, and Spirituality Beyond Earth
2. Awakening the Inner Master: Rosicrucian Secrets to Wisdom and Inner Alchemy
3. Beyond the Surface: Finding Meaning in a Complex World
4. Broken to Blessed: Timeless Biblical Stories of Redemption and Hope
5. Children of the Stars: Humanity's Next Chapter
6. Decoding the Matrix: Finding Freedom in a Fabricated Reality
7. Embrace the Sting: Transforming Envy into Empowerment

8. Golden Dawn: The Alchemist's Guide to Inner Wisdom and Ritual Magic
9. Hidden Hooks: How Stories Snare Our Minds and Move Our World
10. Is Reality Real? Exploring the Mysteries of Physics and Philosophy
11. Masonic Keys: Unlocking Life's Design
12. Mind Unlocked: The CIA's Secret Guide to Shaping Reality
13. Scientology: Core Beliefs, Controversies, and the Quest for Clarity
14. Starborn Legacy: Humanity's Ancient Connection to Interstellar Life
15. Starfed: Alien Wisdom for Vibrant Living
16. Starlit Paths: Lessons from Beyond for Humanity's Rise
17. The Aligned Self: Integrating Ancient Chakra Wisdom into Modern Life
18. The Exorcist's Secret: How to Find Peace in a World of Unseen Conflict
19. The Hidden Kingdom: Jesus' Wisdom for a Transformed Life

20. The New Human: Evolving with Alien Wisdom
21. The Serpent's Gift: How Forbidden Knowledge Shaped Our Destiny
22. The Starseed Path: Embracing Your Cosmic Identity
23. The Viking Spirit: Embracing Life with Courage and Resilience
24. Who Am I, Truly?: Answering the Soul's Deepest Question on the Sufi Path

About the Publisher

This book was published by 22 Lions Publishing.

www.22Lions.com

www.ingramcontent.com/pod-product-compliance
Lightning Source LLC
LaVergne TN
LVHW022305200926
848481LV00004B/1229

* 9 7 9 8 3 3 0 3 0 0 6 4 8 *